active minds

ZOOM INTO SPACE

NEPTUNE

The Stormiest Planet

Ellen Lawrence

sequoia kids media

Photography © NASA, Ruby Tuesday Books, Superstock, Shutterstock 2021 24K-Production; Adelart; Amguy; Balabolka; Dotted Yeti; ideyweb; Marina Yesina

Consultants:

Suzy Gazlay, MA
Recipient, Presidential Award for Excellence in Science Teaching

Kevin Yates
Fellow of the Royal Astronomical Society

Published by Sequoia Kids Media,
an imprint of Sequoia Publishing & Media, LLC

Sequoia Publishing & Media, LLC.,
a division of Phoenix International Publications, Inc.

8501 West Higgins Road
Chicago, Illinois 60631

Customer Service: cs@sequoiakidsbooks.com

www.sequoiakidsmedia.com

ISBN 978-1-64996-207-2

active minds

ZOOM INTO SPACE

NEPTUNE

table of contents

Words shown in **bold** in the text are explained in the glossary.

Welcome to NEPTUNE

Imagine flying to a world that is billions of miles from Earth.

As your spacecraft gets close, you see a blue world covered with clouds.

Inside the clouds are strong winds and fierce storms that could destroy your spacecraft.

You manage to fly through the clouds, but there is nowhere to land.

That's because this faraway world is a huge ball of icy **gases** and liquids.

Welcome to the **planet** Neptune!

No humans have ever visited Neptune, but a spacecraft has. In 1989, a space **probe** called *Voyager 2* flew past the planet.

The Solar System

Neptune is moving through space at over 12,000 miles per hour (19,000 km/h).

It is moving in a huge circle around the Sun.

Neptune is one of eight planets circling the Sun.

The planets are called Mercury, Venus, our home planet Earth, Mars, Jupiter, Saturn, Uranus, and Neptune.

Icy **comets** and large rocks, called **asteroids**, are also moving around the Sun.

Together, the Sun, the planets, and other space objects are called the **solar system**.

Most of the asteroids in the solar system are in a ring called the asteroid belt.

An asteroid

The solar system

Neptune is the farthest planet from the Sun.

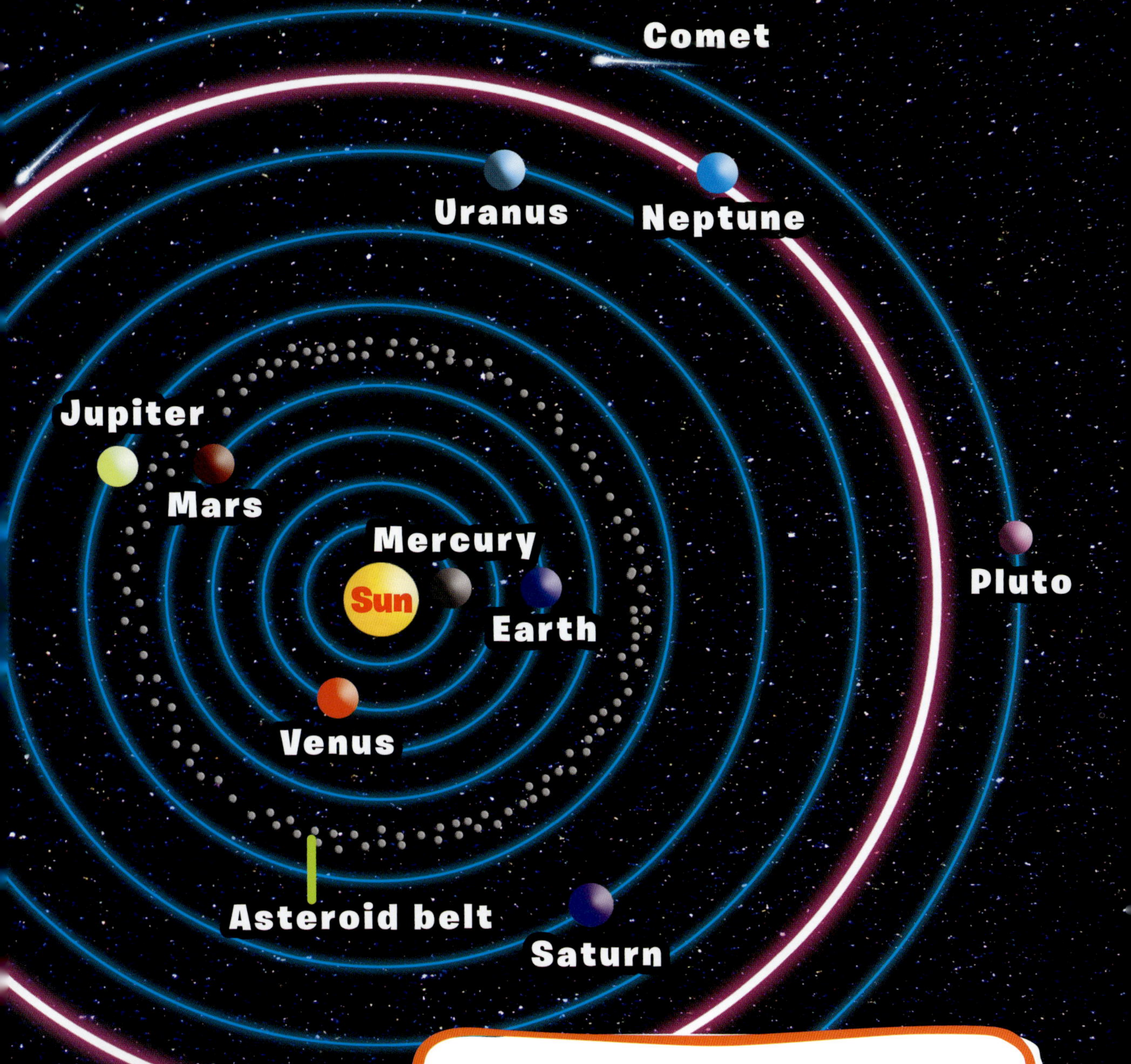

The solar system is home to small planets, called **dwarf planets**. Pluto is a dwarf planet.

Neptune's Amazing Journey

The time it takes a planet to **orbit**, or circle, the Sun once is called its year.

Earth takes just over 365 days to orbit the Sun, so a year on Earth lasts 365 days.

Neptune is farther from the Sun than Earth, so it must make a much longer journey.

It takes Neptune almost 165 Earth years to orbit the Sun.

This means that one year on Neptune lasts for 165 Earth years!

Neptune

As a planet orbits the Sun, it also spins, or **rotates**, like a top.

To orbit the Sun once, Earth makes a journey of about 584 million miles (940 million km). Neptune must make a journey of over 17 billion miles (28 billion km)!

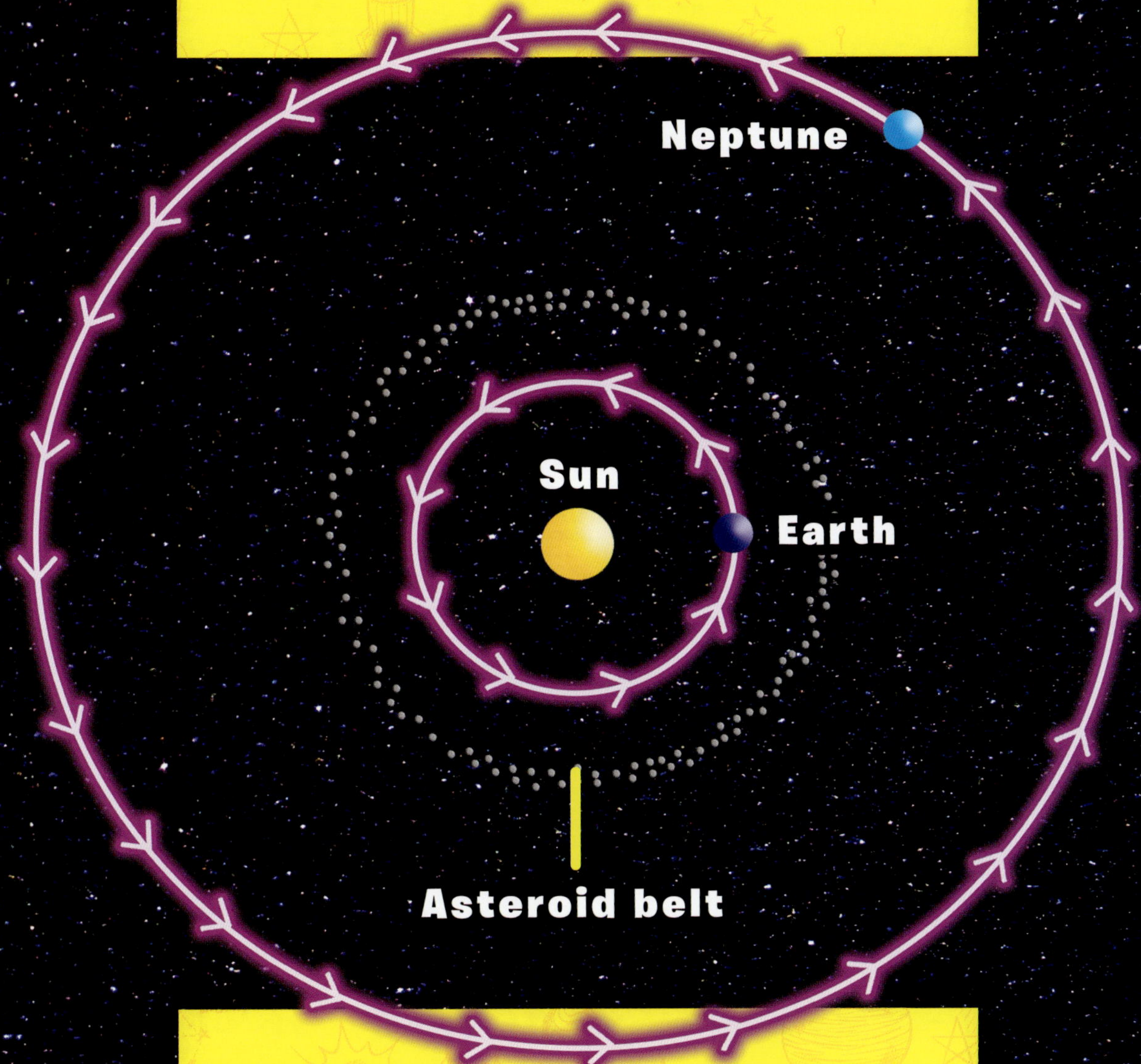

In this picture, we've taken away all the planets in the solar system except for Earth and Neptune. Now it's easy to see Neptune's super-long journey around the Sun!

A Closer Look at NEPTUNE

Neptune is the fourth-largest planet in the solar system.

It is about four times wider than Earth.

Unlike Earth, which is a rocky planet, Neptune doesn't have a solid surface.

The planet is surrounded by a thick layer of gases and clouds called an **atmosphere**.

Beneath its atmosphere, the planet is a huge ball of icy liquids.

The Solar System's Four Largest Planets

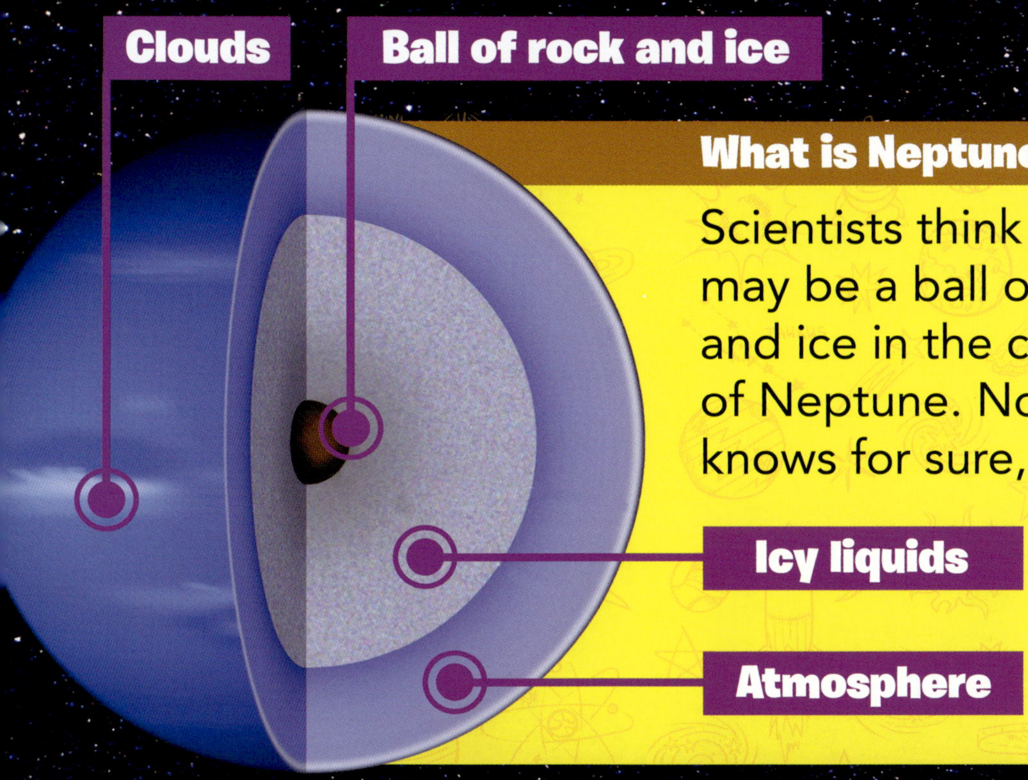

What is Neptune made of?

Scientists think there may be a ball of rock and ice in the center of Neptune. No one knows for sure, though.

Stormy NEPTUNE

Neptune is easily the stormiest planet in the solar system.

Inside its atmosphere there are giant storms with super-fast winds.

The winds can blow at 1,500 miles per hour (2,400 km/h).

These storms are 10 times stronger than the most powerful hurricanes on Earth!

When the *Voyager 2* probe visited Neptune, it took photos of a gigantic hurricane.

The hurricane was called the Great Dark Spot.

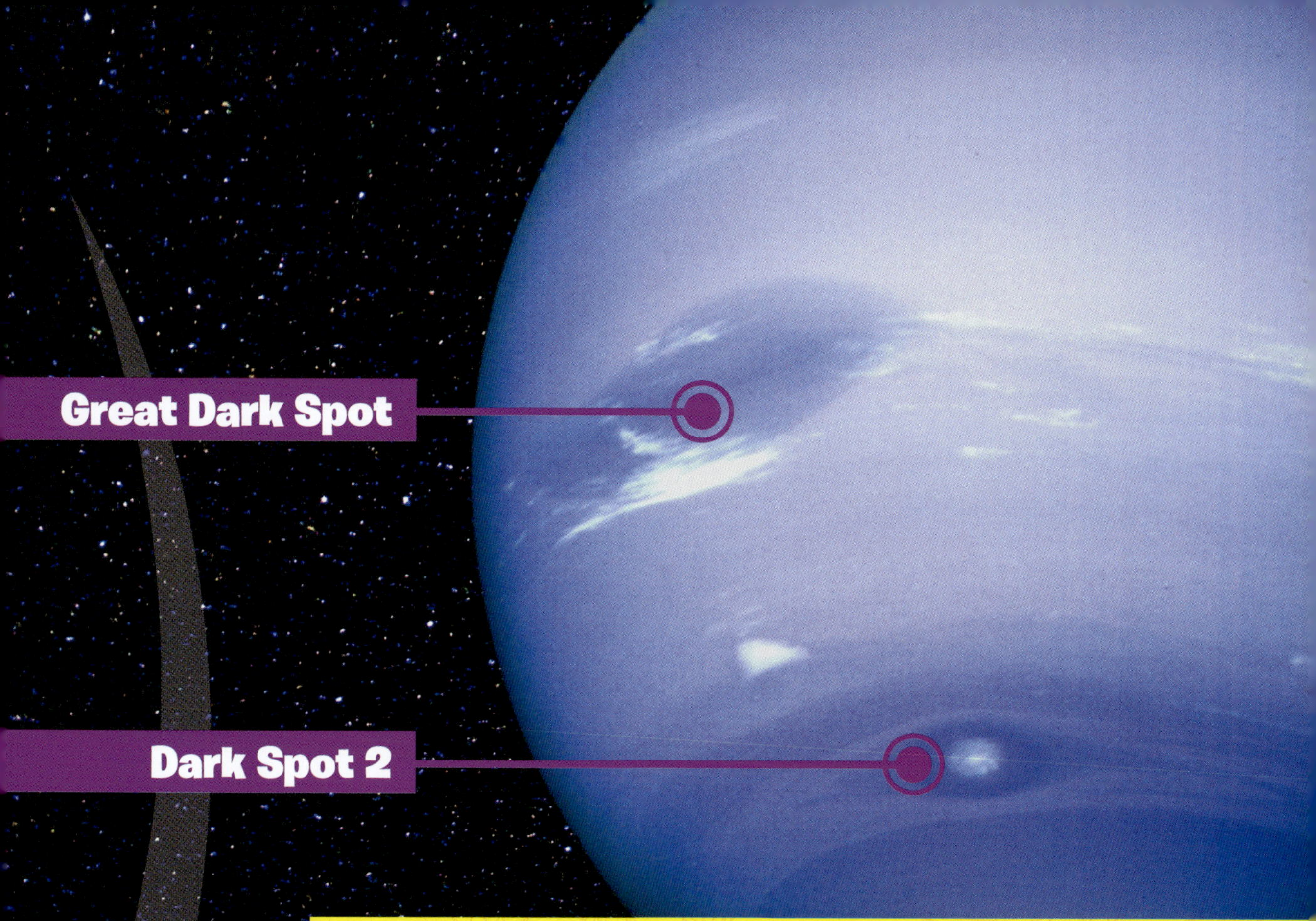

This photo was taken by *Voyager 2*. It shows the Great Dark Spot storm and a smaller storm called Dark Spot 2.

The Great Dark Spot was so big, planet Earth could fit inside it!

NEPTUNE'S

Family of Moons

Neptune has a large family of icy, rocky **moons** orbiting around it.

Our home planet, Earth, has just one moon.

Neptune has at least 14 moons, and scientists think there may be more!

Neptune's largest moon, Triton, is one of the coldest places in the solar system.

This moon has a solid outer crust made of ice.

Beneath the crust and inside the moon, there are icy liquids.

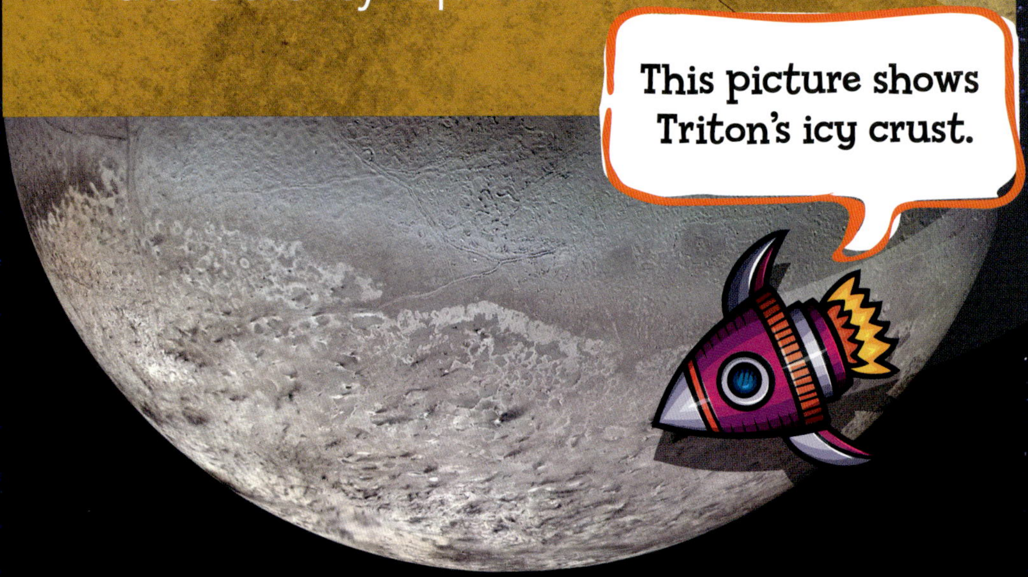

This picture shows what it might look like on Triton. The super-cold moon has ice **volcanoes** on its surface. The volcanoes blast icy liquid from inside Triton up into the sky.

The liquid from Triton's ice volcanoes freezes high above the moon's surface. Then it falls back onto Triton as snow!

A Mission to NEPTUNE

In 1977, the *Voyager 2* space probe blasted off from Earth.

It was on a mission to visit the solar system's most faraway planets.

First *Voyager 2* visited Jupiter, and then Saturn and Uranus.

In 1989, after 12 years in space, it reached Neptune.

Rocket

Voyager 2 blasted off from Earth aboard a rocket. Then it separated from the rocket and flew into space.

To reach Neptune, *Voyager 2* traveled nearly 4.3 billion miles (7 billion km).

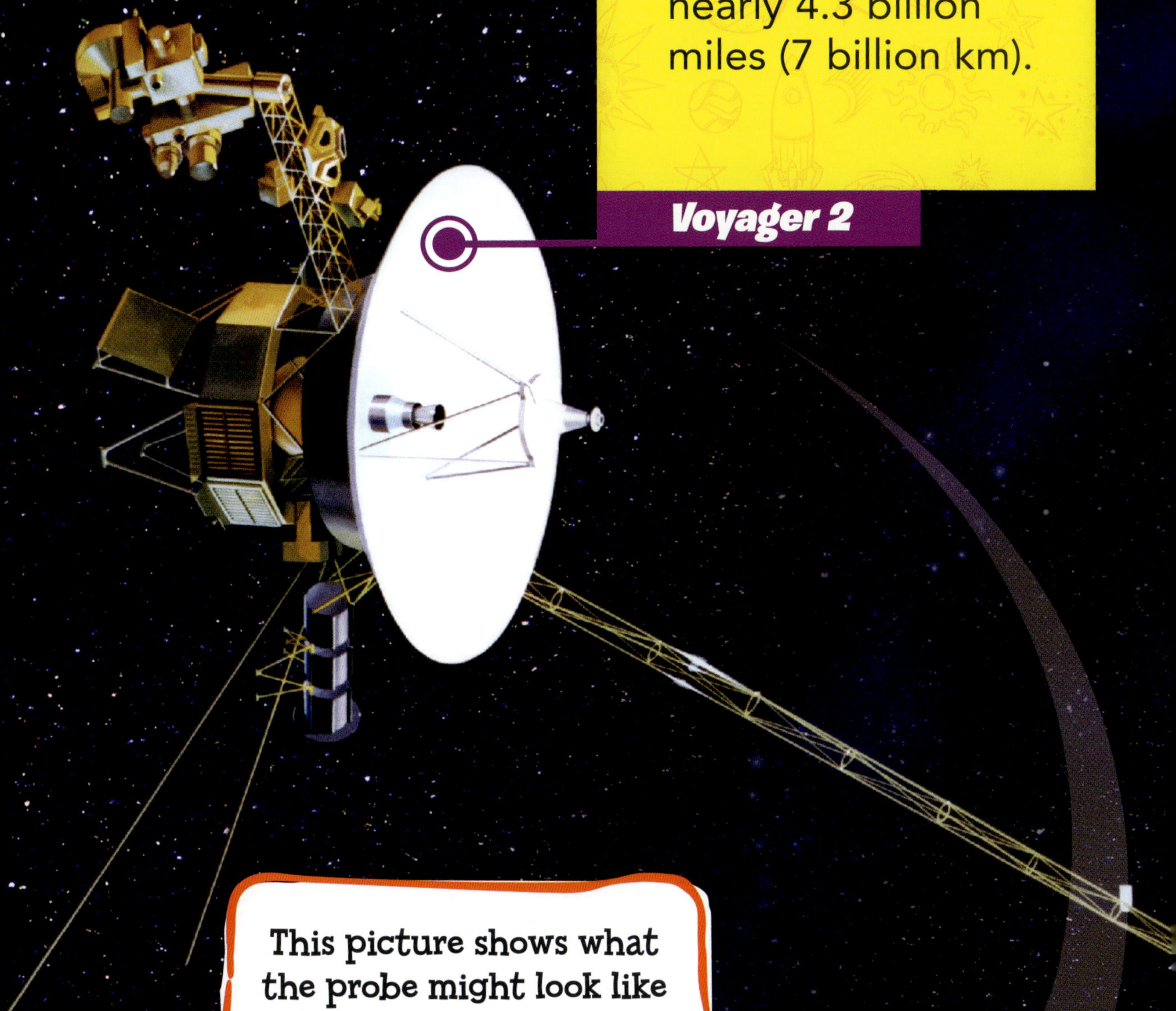

This picture shows what the probe might look like flying through space.

Exciting Discoveries

For many years, scientists thought they saw rings around Neptune through their telescopes.

When *Voyager 2* reached Neptune, it proved they were right.

The probe took photos of rocky, dusty rings circling the planet.

It also discovered five new moons orbiting Neptune.

Voyager 2 is now flying far beyond Neptune.

As of 2021, it is the only space probe to have visited Neptune.

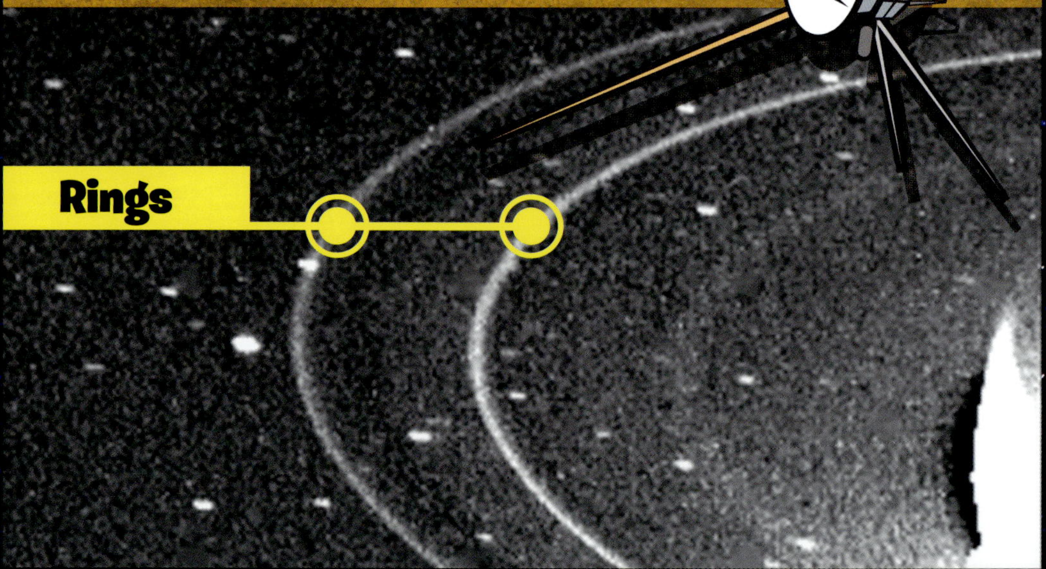

Voyager 2 took hundreds of photos of Neptune. Then it beamed them back to Earth. This is one of the last photos it took as it flew away from the planet.

Scientists think that raindrops made of diamonds fall on Neptune! In 2017, scientists on Earth carried out an experiment. It showed them that methane gas in Neptune's atmosphere turns into solid diamonds. The same thing happens on Uranus.

Neptune Fact File

Here are some key facts about Neptune, the eighth planet from the Sun.

Discovery of Neptune

Neptune was discovered in 1846 by three scientists, Urbain Le Verrier, Johann Galle, and John Couch Adams.

How Neptune got its name

The planet is named after the Roman god of the sea.

Planet sizes

This picture shows the sizes of the solar system's planets compared to each other.

Neptune's size

About 30,599 miles (49,244 km) across

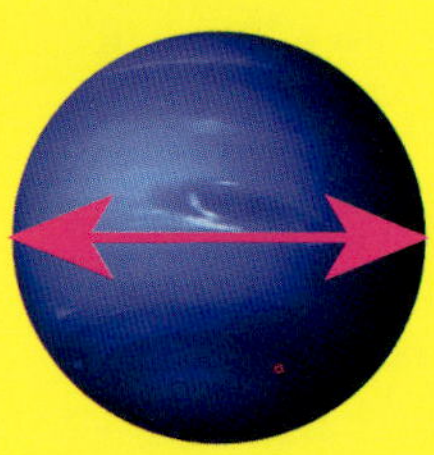

How long it takes for Neptune to rotate once

About 16 Earth hours

Neptune's distance from the Sun

The closest Neptune gets to the Sun is 2,771,162,074 miles (4,459,753,056 km).

The farthest Neptune gets from the Sun is 2,819,185,846 miles (4,537,039,826 km).

Length of Neptune's orbit around the Sun

17,562,271,937 miles (28,263,736,967 km)

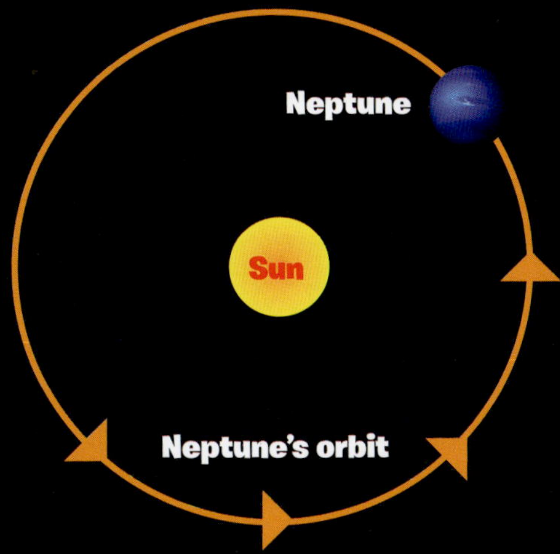

Average speed at which Neptune orbits the Sun

12,158 miles per hour (19,566 km/h)

Length of a year on Neptune

About 60,190 Earth days (about 165 Earth years)

Neptune's moons

Neptune has at least 14 known moons. There are possibly more to be discovered.

Temperature on Neptune

-353°F (-214 °C)

Get Crafty

Melted Wax Crayon Neptune

Using melted wax crayons, make your own picture of the blue planet Neptune and its stormy clouds.

You will need:

- Wax paper
- Scissors
- Blue and purple wax crayons
- A cheese grater
- An iron and an adult helper to use the iron

1. Cut two circles from the wax paper that are about the size of dinner plates.

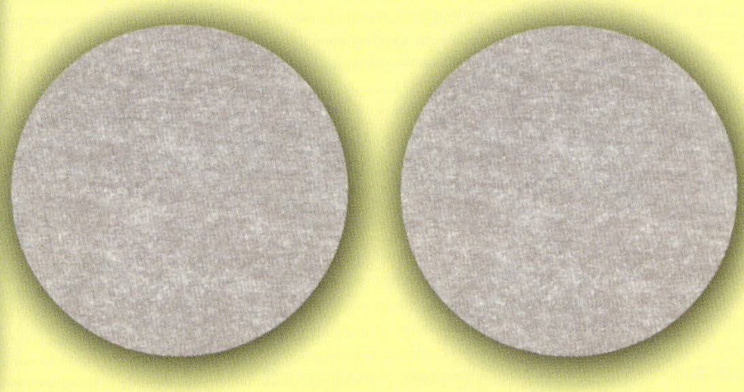

2. Grate the crayons onto one circle of wax paper. Be very careful not to rub your fingers against the grater.

3. Place the other piece of wax paper on top of the grated crayons to make a crayon sandwich.

4. Now ask an adult to iron your crayon sandwich with a cool iron. Keep ironing until the crayons melt and swirl together.

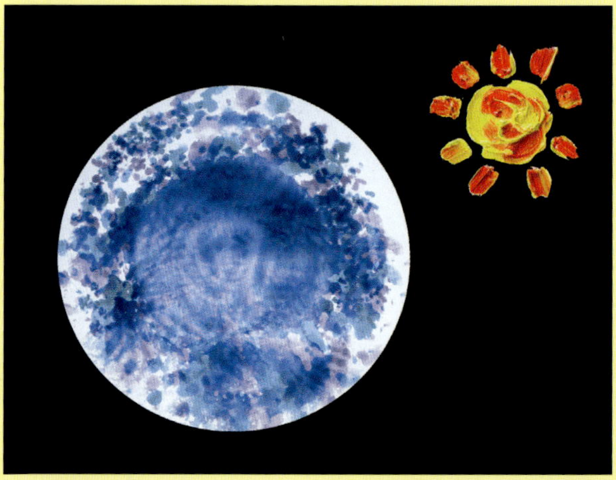

5. Hang your melted crayon Neptune in a window to catch the Sun's light, or use it to make a space picture.

Glossary

asteroid (AS-teh-royd) A large rock that is orbiting the Sun. An asteroid can be as small as a car or bigger than a mountain.

atmosphere (AT-muh-sfeer) A layer of gases around a planet, moon, or star.

comet (KAH-mit) A space object made of ice, rock, and dust that is orbiting the Sun.

dwarf planet (DWARF PLAN-et) A round object in space that is orbiting the Sun. Dwarf planets are much smaller than the eight main planets.

gas (GASS) A substance, such as oxygen or helium, that does not have a definite shape or size.

moon (MOON) An object in space that is orbiting a planet. Moons are usually made of rock, or rock and ice. Some are just a few miles wide. Others are hundreds of miles wide.

orbit (OR-bit) To circle, or move around, another object.

planet (PLAN-et) A large object in space that is orbiting the Sun. Some planets, such as Neptune, are made of gases and liquids. Others, such as Earth, are made of rock.

probe (PROBE) A spacecraft that does not have any people aboard. Probes are usually sent to planets or other objects in space to take photographs and collect information. They are controlled by scientists on Earth.

rotate (ROH-tate) To spin around.

solar system (SOH-ler SIS-tem) The Sun and all the objects that orbit it, such as planets, their moons, asteroids, and comets.

volcano (vol-KAY-noh) An opening in the surface of a planet or moon through which liquids and gases erupt. On Earth, the liquids are hot, melted rock. On cold, distant moons, they can be icy liquids, such as water.

Index

Read More

Explore Neptune
Lightning Bolt Books — Planet Explorer
Jackie Golusky
Lerner Publishing Group
2021

Neptune
Planetary Exploration
Greg Roza
Rosen Publishing
2017

Bonus Downloadable Activities!

www.sequoiakidsmedia.com